Exercises and Applications to accompany

Food for Fifty

Thirteenth Edition

Katrina Warner

Prentice Hall

Boston Columbus Indianapolis New York San Francisco Upper Saddle River
Amsterdam Cape Town Dubai London Madrid Milan Munich Paris Montreal Toronto
Delhi Mexico City Sao Paulo Sydney Hong Kong Seoul Singapore Taipei Tokyo

Editor in Chief: Vernon Anthony
Acquisitions Editor: William Lawrensen
Editorial Assistant: Lara Dimmick
Director of Marketing: David Gesell
Marketing Manager: Kara Clark
Senior Marketing Coordinator: Alicia Wozniak
Marketing Assistant: Les Roberts
Associate Managing Editor: Alexandrina Benedicto Wolf
Project Manager: Kris Roach
Senior Operations Supervisor: Pat Tonneman
Operations Specialist: Deidra Skahill
Senior Art Director: Diane Ernsberger
Cover Designer: Ilze Lemesis
Cover Art: iStock
Compositor: Kelly Barber
Printer/Binder: Demand Production Center
Cover Printer: Demand Production Center
Text Font: Times Roman

Copyright © 2011 by Pearson Education, Inc., publishing as Prentice Hall, Upper Saddle River, New Jersey, 07458. All rights reserved. Manufactured in the United States of America. This publication is protected by Copyright, and permission should be obtained from the publisher prior to any prohibited reproduction, storage in a retrieval system, or transmission in any form or by any means, electronic, mechanical, photocopying, recording, or likewise. To obtain permission(s) to use material from this work, please submit a written request to Pearson Education, Inc., Permissions Department, 1 Lake Street, Upper Saddle River, New Jersey, 07458.

Prentice Hall
is an imprint of

www.pearsonhighered.com

10 9 8 7 6 5 4 3 2 1
ISBN 10: 0-13-507225-5
ISBN 13: 978-0-13-507225-7

Table of Contents
Workbook for *Food for Fifty*, Molt; Prepared by Katrina E. Warner

Factor method, pg 11 1

Percentage method, pg 11 3

Abbreviations used in recipes, pg 7 5

Direct-reading tables, tables 2.1 and 2.2, pg 16 and 18 7

Direct-reading table for increasing home-sized recipes,
    table 2.3, pg 24 9

AP/EP concepts 11

Yield relationships 13

Amounts of food to serve 50, table 4.1, pg 61 15

Food weights and approximate equivalents in measure,
    table 4.2, pg 71 17

Ingredient substitutions, table 4.5, pg 88 18

Ingredient proportions, table 4.6, pg 90 19

Ounces and decimal equivalents of a pound and grams,
    table 4.7, pg 90 21

Weight and approximate measure equivalents for commonly
    used foods, table 4.10, pg 92 23

Common can sizes, table 4.11, pg 94 25

Drained weight 27

Metric equivalents for weight, measure, and temperature,
table 4.12, pg 94 29

Convection oven baking times and temperatures, table 6.3, pg 154 31

Guidelines for reducing the risk of food-borne illness,
table 7.1, pg 163 33

Cold food storage temperatures, table 7.3, pg 164 35

Safe internal temperatures for cooked foods, table 7.6, pg 165 37

Food serving temperatures and holding times, table 7.7, pg 166 39

Fresh herb descriptions, flavor, and usage, table 5.18, pg 128 41

Salt and pepper seasonings, table 5.23, pg 139 43

Pan capacities for baked products, table 7.16, pg 184 44

Dipper equivalents, table 7.18, pg 186 45

Ladle equivalents, table 7.19, pg 186 47

Recommended mixer bowl and steam-jacketed kettle sizes
for selected products, table 7.20, pg 187 49

Basic proportions and yields for converted rice, table 16.1, pg 506 51

Create a production worksheet, pg 144 53

Mise en place steps 55

Using menu planning principles, plan a 4-day menu 57

Strategies to reduce fat, sodium, and sugar 59

Using cheeseburger pie recipe, calculate amount of
AP ground beef needed 61

Prepare an oven schedule for preparing a menu with
four oven cooked items 63

Cost per pound AP and EP 65

Buffet set up 67

Menu for all religions 69

Name_____   Date_____

Recipe Adjustment: Factor Method, pg 11 in FFF

Using the recipe on pg 365 in FFF, Chocolate Pudding, adjust to the following servings:
Chocolate Pudding
Desired yield: 130 servings, ½ cup portion   Current yield: 50 servings, ½ cup portion

Ingredients	Recipe Amount 50 portions	Converted quantities	Factor	Extended amount (Calculated amount)	Extended amount (Common measure)

Chocolate Pudding
Desired yield: 30 servings, ½ cup portion   Current yield: 50 servings, ½ cup portion

Ingredients	Recipe Amount 50 portions	Converted quantities	Factor	Extended amount (Calculated amount)	Extended amount (Common measure)

Name_____   Date_____

Recipe Adjustment: Percentage Method, pg 11 in FFF

Using the recipe on pg 233 in FFF adjust to the following servings:
Basic Muffins Desired yield: 150 portions   Current yield: 50 servings, 2¼ oz portion

Ingredients	Recipe Amount 50 portions	Converted quantities	Percentage	Calculate weights	Convert to pounds and oz

Basic Muffins Desired yield: 25 portions   Current yield: 50 servings, 2¼ oz portion

Ingredients	Recipe Amount 50 portions	Converted quantities	Percentage	Calculate weights	Convert to pounds and oz

Name_____ Date_____

Abbreviations Used in Recipes – Worksheet

Directions: Working individually, write the abbreviations for the following words that are often used in recipes.

Words: **Abbreviations:**

As Purchased   _____
Cup   _____
Degrees Celsius   _____
Degree Fahrenheit   _____
Fluid ounce   _____
Gallon   _____
Gram   _____
Kilogram   _____
Liter   _____
Milliliter   _____
Number   _____
Ounce   _____
Package   _____
Pint   _____
Pound   _____
Quart   _____
Tablespoon   _____
Teaspoon   _____
Weight   _____
Edible Portion   _____

Name_____   Date_____

Table 2.1 Direct-reading table for adjusting weight ingredients of recipes divisible by 25

Table 2.2 Direct-reading table for adjusting recipes with ingredient amounts given in volume measurement and divisible by 25

Using the recipe for Deviled Eggs, pg 395, use tables 2.1 and 2.2 to quickly convert the recipe:

1. Decrease recipe to serve 25 portions

Ingredient	Original amount	25 portions

2. Increase recipe to serve 400 portions

Ingredient	Original amount	400 portions

Name_____   Date_____

Table 2.3 Direct-reading table for increasing home-sized recipes with ingredient amounts given in volume measurement and divisible by 8

Use the Tarragon chicken home-recipe below to adjust servings as indicated:

Ingredient	Original amt for 8 portions	24 portions
Chicken breasts, boneless	3 lb	
Tarragon, fresh, chopped	1/3 c + 5 ¼ tsp	
Fresh lemon juice	3 Tbsp + 1 ¾ tsp	
Olive oil	½ cup	
Salt, sea	¼ tsp (s)	
Pepper, black ground	½ tsp (s)	

Ingredient	Original amt for 8 portions	64 portions
Chicken breasts, boneless	3 lb	
Tarragon, fresh, chopped	1/3 c + 5 ¼ tsp	
Fresh lemon juice	3 Tbsp + 1 ¾ tsp	
Olive oil	½ cup	
Salt, sea	¼ tsp (s)	
Pepper, black ground	½ tsp (s)	

Name_____   Date_____

Yield Concepts

A. Yield Terminology

B. **As Purchased (AP)** – the total product purchased or received (includes waste)
   1. <u>Examples</u>: block of cheese, bag of flour, raw strip steak, precooked pot roast, whole apple, unpeeled carrot, whole cantaloupe, whole bunch of celery, frozen broccoli

C. **Edible Portion (EP)** – the amount of food, usually produce (fresh fruits or vegetables) remaining after all waste (inedible) parts have been removed.

   Must be calculated for all FRESH fruits and vegetables used in recipes so that yield and cost information is correct.
   1. <u>Examples</u>: a cored, peeled apple, a peeled and trimmed carrot, cantaloupe wedges, trimmed celery stalk, diced onion

D. As Purchased : Edible Portion Relationships
   1. <u>To determine % yield</u> = final weight / original weight **or** EP / AP

   2. <u>To determine the EP</u>: EP = AP x % yield

   3. <u>To determine the AP</u>: AP = EP / % yield

   4. Approximate yield for prepared fruits and vegetables:

Item	% yield
Apples	.78
Bananas	.65
Carrots	.70
Celery	.83
Cucumbers	.84
Honeydew	.46

Item	% yield
Lettuce, head	.76
Mushrooms	.98
Onions	.88
Parsley	.92
Peppers, green	.80
Pineapple	.54

Name_____   Date_____

E. **Exercises** – Use the values above to complete the following exercises.

(1) Calculate AP amounts for the following EP fruit and vegetable amounts.
Remember: AP = EP / % yield

a. 200 lb onions	b. 25.4 lb lettuce
c. 8 lb 4 oz carrots	d. 90 lb celery
e. 215 lb bananas	f. 15.6 lb apples
g. 17.3 lb pineapple	h. 18.6 lb peppers
i. 50 lb 2 oz cucumbers	j. 400 lb honeydew
k. 10 lb mushrooms	l. 2.5 lb parsley

(2) Calculate EP amounts for the following AP fruit and vegetable amounts.
Remember: EP = AP x % yield

a. 200 lb onions	b. 25.4 lb lettuce
c. 8 lb 4 oz carrots	d. 90 lb celery
e. 215 lb bananas	f. 15.6 lb apples
g. 17.3 lb pineapple	h. 18.6 lb peppers
i. 50 lb 2 oz cucumbers	j. 400 lb honeydew
k. 10 lb mushrooms	l. 2.5 lb parsley

Name_____ Date_____

Yield

1. Relationship between final and original weight

2. Amount of product remaining after production (prep, paring, handling, cooking, draining) or portioning

3. Yield definitions:
    a. Expressed as a percent (%)
    b. Yield % = final weight / original weight
    c. Yield % = 100% - % loss

4. Exercise: Calculate the yield for the following: Answers:

1. A case of lettuce weighs 48.75 pounds (without the box). The yield for cleaned lettuce is 77%. How much lettuce can be put in the tossed salad?	
2. A roast weighing 152 lb 14 oz was roasted in a convection oven. The cooked roast weight was 102.5 lb. What is the yield?	
3. Chef cooked 200 lb of raw spaghetti. After cooking, the spaghetti weighed 300 lb. What is the yield?	
4. A #10 can of pears weighs 6 lb 6 oz. After draining the juice, Jim weighed the peaches again and found he had 4.75 lb. What is the yield?	
5. A #10 can of green peas weighs 6 lb 3 oz. If 35% of the can is liquid, what is the weight of the drained green peas?	
6. A fast-food 4 oz ground beef patty shrinks 33% during cooking. What is the cooked weight of the beef patty?	

Name_____   Date_____

Table 4.1 – Amounts of Food to Serve, Yield, and Food Equivalent Information

Use table 4.1 to calculate the amount of AP product needing to be purchased to cater a party of 225

Ingredient	Amt for 50	AP Yield %	Factor	Amount to purchase
Ground beef (73% lean) 4 oz EP patty				
Catsup 1 oz portion				
Salad bulky veg 1c				
Lettuce leaf for garnish 1 leaf				
Potatoes French fried 4 oz portion				
Cheese sand. Slices 1 oz each				
Bread 1 ½ lb loaf 2 slices portion				

Use table 4.1 to calculate the amount of AP product needing to be purchased to cater a party of 30

Ingredient	Amt for 50	AP Yield %	Factor	Amount to purchase
Ground beef (73% lean) 4 oz EP patty				
Catsup 1 oz portion				
Salad bulky veg 1c				
Lettuce leaf for garnish 1 leaf				
Potatoes French fried 4 oz portion				
Cheese sand. Slices 1 oz each				
Bread 1 ½ lb loaf 2 slices portion				

Name_____ Date_____

Volume / Weight Relationships

A. Quantity food service measurement procedures

   1. Weigh solids (except those that are less than one ounce – unless you have a scale that can weigh that amount)
   2. Measure liquids (unless recipe calls for weight)
   3. WHY?
       a. accuracy
       b. computer applications
       c. costing

B. To convert ingredients from volume to weight:

   1. Use table 4.10 Food Weight and Approximate Equivalents (pg 92) in FFF for basic information and charts starting on pg 71.
   2. Weight (ingredient) = (Volume (recipe) X Weight (table 4.10)) / Volume (table)
   3. EXERCISE: Volume / Weight Conversions

(1) A recipe calls for 9 Tbsp baking soda. Convert this to weight.	
(2) A recipe calls for 32 cups of chopped onion. What is the weight of the onion?	
(3) What is the weight of 35 cups of sliced, fresh mushrooms?	
(4) What is the weight of 10 Tbsp of ground dry mustard?	
(5) How many (by count) miniature marshmallows are there in 2 lb?	

Name_____   Date_____

Table 4.5 Ingredient substitutions

1. Using the recipe on page 332, Fudge brownies, calculate a substitution for the baking powder.

2. Using the recipe on pg 231, Baking powder biscuits with the buttermilk variation at the bottom of the page, calculate a substitution for the buttermilk if you only had whole milk.

3. Using the recipe on pg 652, Garlic herb dressing, calculate how much garlic, minced, dry you would need to substitute in the recipe if it was doubled.

4. Using the recipe on pg 571, Spinach lasagna, calculate how much fresh basil and oregano you need for the recipe to substitute the dried herbs.

5. Using the recipe on pg 468, Glazed baked ham with the honey mustard glaze variation, calculate the substitution you would need for the honey.

Name_____   Date_____

Table 4.6 Ingredient proportions

1. How much salt would you add to a 25 lb roast of beef?

   _____

2. If a recipe for pancakes calls for 5 lb of flour, how much baking powder would you need?

   _____

3. For a medium béchamel (white) sauce, how much flour would you need for 6 gallons of milk?

   _____

4. You are making a raspberry gelatin dessert using 3.5 gallons of liquid. How much gelatin do you need?

   _____

5. In making biscuits using 2qt + 1pt, how much flour will you need?

   _____

Name_____   Date_____

Table 4.7 Ounces and decimal equivalents of a pound and grams

Using table 4.7, convert the following recipe

Ingredient	Original amount	Convert to pounds	5 times the recipe	Extended amount in pounds
Rice, Arborio	2 lb 4 oz			
Water	8 lb 8.5 oz			
Olive oil	3 oz			
Butter	12 oz			
Parmesan cheese	14 oz			

Name_____   Date_____

Table 4.10 Weight (1-16 oz) and approximate measure equivalents for commonly used foods

Using the recipe for Brownies, pg 371, convert weights to measure or vice versa

Ingredients	In recipe as weight or measure	Converted to weight or measure
Eggs		
Sugar, granulated		
Shortening, melted		
Margarine, melted		
Flour, cake		
Baking powder		
Salt		
Nuts, chopped		

Name_____   Date_____

Table 4.11 Common can sizes

1. Using the recipe on pg 691, Italian tomato sauce, calculate the size and number of cans you will need to purchase for the following ingredients if you did the recipe and then four times the recipe.

	Original recipe	4 times the recipe
a. Tomato juice	_____	_____
b. Tomato puree	_____	_____
c. Tomato paste	_____	_____

2. Using the recipe on pg 598, Garbanzo bean salad, calculate the size and number of cans you will need to purchase for the following ingredients if you did the original recipe and then doubled it.

	Original recipe	2 times the recipe
a. Garbanzo beans	_____	_____
b. Red beans	_____	_____
c. Pinto beans	_____	_____
d. Black olives	_____	_____

Name_____   Date_____

Drained Weight:

1. Definition
    a. Quantity of a canned fruit or vegetable remaining after the item has drained for two minutes.
    b. Drained weight = Total Contents (Can) Weight – Drained Liquid Weight
    c. Drained weight % = Drained Weight / Total Contents Weight
    d. Drained weights and can volumes vary for each fruit and vegetable
    e. For classroom purposes, we will use the following definitions:
        (1) The net contents of a #10 can is 6 lb 8 oz or 6.5 lb
        (2) The drained weight percent is 65%
        (3) The volume of a #10 can is 3 qt
        (4) A case has 6 #10 cans in it

2. EXERCISE: Calculate the drained weight:

(1) What is the drained weight of a can of green beans?	
(2) What is the drained weight of a case of cut corn?	
(3) What is the volume of crushed pineapple from a #10 can?	
(4) What is the volume of 4 #10 cans of ketchup?	
(5) What is the weight of a case of diced tomatoes in #10 cans?	

Name_____   Date_____

Table 4.12 Metric equivalents for weight, measure, and temperature

Convert the following ingredients from Brownies, pg 331 into metric

Ingredients	In recipe	Converted to metric
Sugar, granulated		
Shortening, melted		
Margarine, melted		
Vanilla		
Flour, cake		
Baking powder		
Salt		
Nuts, chopped		
Oven temperature		

Name_____ Date_____

Table 6.3 Convection oven baking times and temperatures

1. Using the recipe on page 484, Oven-fried chicken, adjust the recipe procedures to reflect using a convection oven.

2. Using the recipe on page 464, Breaded pork chops, adjust the recipe procedures to reflect using a convection oven.

3. Using the recipe on page 440, Meat loaf, adjust the recipe procedures to reflect using a convection oven.

Name_____   Date_____

Table 7.1 Guidelines for reducing the risk of food-borne illness

1. List three things to look for when purchasing/receiving food:
    a. _____
    b. _____
    c. _____

2. How should food be stored to prevent cross-contamination?
    _____
    _____
    _____

3. List three things a foodservice worker should do to prevent food-borne illness while preparing food:
    a. _____
    b. _____
    c. _____

4. What procedure can a foodservice worker use to know if a cooked food is done (example: baked chicken)?
    _____
    _____

5. List four guidelines a foodservice worker should use while serving food:
    a. _____
    b. _____
    c. _____
    d. _____

Name_____   Date_____

Table 7.3 Cold food storage temperatures

1. Which items have the shortest refrigerator shelf life according to the chart?

   _____

2. Which items have the longest refrigerator shelf life? Why do you think that is?

   _____

3. Which items have the lowest refrigerator storing temperatures?

   _____

4. If a restaurant received ground beef on Monday for chili but didn't need it until Saturday, what could they do to deter bacterial growth without freezing it?

   _____

Name_____ Date_____

Table 7.6 Safe internal temperatures for cooked foods

List the following safe internal temperatures for the following foods (all are from recipes in book):

1. _____ Chicken quesadilla

2. _____ Glazed baked ham

3. _____ Swedish meatballs

4. _____ Broiled tuna with white beans

5. _____ Salmon loaf

6. _____ Cheeseburger pie

7. _____ Quiche

8. _____ Chicken tetrazzini

9. _____ Cheese soufflé

10. _____ Baked ziti with four cheeses from last night

Name_____   Date_____

Table 7.7 Food serving temperatures and holding times

Planning a breakfast buffet, the hostess would like the food to be available to her guests for 1 hour.

Help her plan the timing of the breakfast with the following foods:

Buffet item	Plan of action
Orange juice in pitchers	
Coffee/decaf	
Danish/muffins/donuts/biscuits	
Egg casserole	
Eggs, scrambled	
Ham slices	
Sausage gravy	

Name_____ Date_____

Table 5.18 Fresh herb descriptions, flavors, and usage

1. What ingredients could you use to add a licorice or anise flavor to food?

   _____

2. Which herb has a close flavor to oregano?

   _____

3. Which herbs have a lemon or citrus flavor?

   _____

4. Which herbs can be used with poultry dishes?

   _____

5. Which herbs are sour or pungent?

   _____

Name_____   Date_____

Table 5.23 Salt and pepper seasonings

1. Is salt an herb, spice, or seasoning?

   _____

2. Which salt is lighter in weight and flakier than table salt?

   _____

3. Which salt has a chemical added to it to keep it free flowing?

   _____

4. Which peppercorns are not from the same vine as other pepper?

   _____

5. Which peppercorns are unripe and either freeze-dried or put in a brine or vinegar to stay soft?

   _____

6. How are black pepper and white pepper different?

   _____

7. Which salt has a gray-brown color and why?

   _____

8. What is the recommended mixture of salt and pepper to season foods?

   _____

Name_____   Date_____

Table 7.16 Pan capacities for baked products

1. Using the Yellow cake recipe on page 294, how many 9-inch round cakes will it make?

   _____

2. Using the Yellow angel food (sponge) cake recipe on page 289, how many sheet pans (full or half) will it make?

   _____

Table 7.17 Hotel/counter pan capacities

1. How many half-size, 2.5-inch pans will the recipe on page 530, Fresh tomato linguine need?

   _____

2. Recipe on page 536, spaghetti with meatballs – how many and what size pans would you need to serve on a buffet?

   _____

Name_____   Date_____

Table 7.18 Dipper equivalents

1. If using chicken salad, pg 624 as an entrée, how many number 6 dipper portions will you have?

   _____

2. How many cookies will the recipe for chocolate chip cookies, pg 322, make if using a number 60 dipper?

   _____

3. How many hamburger patties can you portion from a 10 lb tube of beef using a number 10 dipper?

   _____

4. Using a number 12 dipper, how many scoops of ice cream will you get from a 5 gallon container?

   _____

5. How many cupcakes can you get from the yellow cake recipe, pg 294, using a #20 dipper?

   _____

Name_____ Date_____

Table 7.19 Ladle equivalents

1. How many 6 oz servings will you get from Broccoli and cheese soup, pg 759?

   _____

2. If serving bacon dressing, pg 643, to 50 people, could you use a 2 oz ladle? Why or why not?

   _____

3. If serving 100 people a tasting of beef stew, pg 444, what is the largest ladle you could use?

   _____

4. After serving the 100 people, how much would be left over?

   _____

   What size ladle would you use to serve Barbecue sauce (cooked), pg 688, to 150 people?
   How much would be left over?

   _____

Name_____ Date_____

Table 7.20 Recommended mixer bowl and steam-jacketed kettle sizes for selected products

1. What size mixer bowl would you need to make 10 dozen jumbo chunk chocolate cookies, pg 324?

   _____

2. What size mixer bowl would you need to make 100 portions of yellow cake, pg 294?

   _____

3. What size mixer bowl would you need to make 200 portions of egg and sausage bake, pg 391?

   _____

4. What size kettle would you need to cook the pasta for 150 portions for spaghetti with meat sauce, pg 536?

   _____

5. What size kettle would you need to cook the sauce for 150 portions for the same recipe in #4?

   _____

Name_____   Date_____

Table 16.1 Basic proportions and yields for converted rice

1. How much raw rice will you need to end up with 10 4 oz cooked servings?

   _____

2. How many 3 oz cooked servings will you get with 10 lb raw rice?

   _____

3. How much raw rice will you need to make cooking rice, pg 542 if the portion size increases to 5 oz for 50 servings?

   _____

4. How much liquid will you need for 9 lb raw rice?

   _____

Name_____   Date_____

Create a production worksheet, pg 144

Use the following recipes for a meal period that is estimated to serve 125:
Macaroni salad, pg 607
Chicken pot pie, pg 495
Baked apples, pg 375

Employee name	Item to prep	Prep procedure	EP prep amount	AP amount needed	Start time	End time/ total time
Example: Joe	*Onions*	*Chopped*	*5 oz for entrée 3 lb 2 oz for salad*	*55oz/.88 = 62.5oz (used table 4.1) 62.5oz/16 = 3# 14.5oz*	*9am*	*9:30am/ 30 minutes*

Name_____ Date_____

Learning outcomes:

- ***The student will learn to group production steps together that increase labor efficiency.***

- ***The student will learn to schedule the production of food items sequentially so as to assure quality food.***

- ***The student will learn the benefits of using a production sheet as a communication tool.***

- ***The student will identify production controls.***

Name_____   Date_____

Mise En Place steps

Identify the mise en place tasks appropriate for 175 servings of Vegetable and Tofu Jambalaya (p. 591) served at the lunch meal.

_____

_____

_____

_____

_____

_____

_____

_____

_____

_____

Learning outcomes:

- *The student will learn to identify the sequence of steps to produce a product.*
- *The student will learn to evaluate the quality food impact of pre-preparation tasks.*
- *The student will learn how to use labor efficiently.*

Name_____   Date_____

Using the menu planning principles, plan a 4-day menu with the following guidelines per day

1,800 calories with 55% carbohydrates, 15% protein and 30% fat

| | **Breakfast** | **Lunch** | **Dinner** | **Snack(s)** |

1. _____

2. _____

3. _____

Name_____ Date_____

4. _____

   _____

   _____

   _____

Learning outcome:

- ***The student will learn to plan a menu with guidelines using principals listed on pages 33-34.***

Name_____ Date_____

Using the suggestions listed on pages 12 – 14, list strategies to reduce fat/sodium/sugar in the following recipes and calculate the outcome of those reductions:

1. Pancakes, pg 252_____

   _____

   _____

   _____

2. Broccoli and cheese soup, pg 759_____

   _____

   _____

   _____

3. Sour cream dressing, pg 643_____

   _____

   _____

   _____

Name_____   Date_____

4. Nut bread, pg 247_____

_____

_____

_____

Learning outcomes:

- *The student will demonstrate the reduction of fat, sodium, and sugar in recipes.*
- *The student will demonstrate the outcome in the recipes of the reduction strategies.*
- *The student will do these without sacrificing the outcome of the recipe in regard to quality and taste.*

Name_____ Date_____

In the Cheeseburger pie recipe, pg 457, there is a 66% yield on the ground beef. If using the same 8 lb EP beef amount in the recipe but with the following fat contents (found on page 65), calculate the amount of AP ground beef needed and with the prices listed, which would be most economical?

1. Ground, $1.99/lb:_____

   _____

   _____

   _____

2. Lean, $2.49/lb:_____

   _____

   _____

   _____

3. Extra Lean, $2.99/lb:_____

   _____

   _____

   _____

Name_____   Date_____

Learning outcomes:

- ***The student will learn the calculation to find EP with different yield values.***
- ***The student will learn the cost difference between fat contents of ground beef.***

Name_____   Date_____

Prepare an oven schedule for preparing a menu with four oven cooked items:

Whole Turkey (pgs. 485 - 486), Cornbread Dressing (p. 244), Bread Dressing (p. 504), and Pumpkin Cake Roll (p. 309). Note oven temperatures and cooking times.

Data to use:

1. One oven available. Size (18" deep, 26" wide, 14" tall).

2. Noon meal.

3. Forecast 25 people.

4. No other items will fit in the oven with the turkey.

How many and what total weight of whole turkeys would you order?

Oven Schedule	TEMPERATURE	TIME IN	TIME OUT
Turkey			
Cornbread Dressing			
Bread Dressing			
Pumpkin Cake Roll			

Name_____   Date_____

Learning outcomes:

- *The student will learn how to utilize oven space.*
- *The student will learn how to plan a meal for 25 in a certain time period.*
- *The student will learn to determine AP/EP needed for raw turkey.*

Name_____   Date_____

Cost per pound AP and EP:
Use table 4.1 for help with EP calculations

1. Potatoes are $4.39 for a 5 lb bag, what is the AP price per pound?

   _____

   What is the EP price per pound?

   _____

2. Tomatoes are $6.39 for 3 lb, what is the AP price per pound?

   _____

   What is the EP price per pound for peeled & seeded?

   _____

3. Bulk sausage is $5.96 for 3 lb, what is the AP price per pound?

   _____

   What is the EP price per pound for cooked lean?

   _____

4. Macaroni is $.99 per pound, how much is a cooked 4 oz portion?

   _____

Learning outcome:

- ***The student will learn how to determine the AP and EP prices for items.***

Name_____   Date_____

Using the information on pages 47- 52, design a buffet set up for both 100 people and 40 people for the following menu:

Barbecued spareribs, pg 466
Vegetarian spaghetti, pg 533
Coffee/decaf
Bowl of salad greens, 3 sides to go with – your choice
Butterscotch pecan cookies, pg 322
Iced tea
Applesauce, pg 377
2 salad dressings – your choice
Fruit punch
Ranch style beans, pg 777
Rice pilaf, pg 554
Seasoned whole kernel corn, pg 787
Peanut butter cake, pg 301
Basic roll dough, pg 272
Vanilla bean ice cream

_____

_____

Learning outcome:

- *The student will learn the process of planning a buffet with proper flow and size according to the items being offered and amount of guests.*

Name_____   Date_____

According to the food practices of different religions, table 3.3, pg 32, create a day's menu that would satisfy ALL groups.

Breakfast:
    Entrée: _____

    Two side dishes/accompaniments: _____

    _____

    Beverages: _____

Lunch:
    Appetizer: _____

    Entrée: _____

    Two side dishes/accompaniments: _____

    _____

    Dessert: _____

    Beverages: _____

Dinner:
    Appetizer: _____

    Entrée: _____

    Two side dishes/accompaniments: _____

    _____

    Dessert: _____

    Beverages: _____

Learning outcome:
- ***The student will learn to plan a menu according to dietary needs/beliefs.***